MW01599330

201207

BROWN BEARS

Published by Smart Apple Media
1980 Lookout Drive, North Mankato, Minnesota 56003

Design and Production by The Design Lab/Kathy Petelinsek

Photographs by Lynn M. Stone

Library of Congress Cataloging-in-Publication Data
Gish, Melissa.
Brown bears / by Melissa Gish
p. cm. — (Northern Trek)
Includes resources, glossary, and index
Summary: Describes the behavior, habitat, physical characteristics, endangered status,
and conservation of Alaskan brown bears, also known as Kodiak bears.
ISBN 1-58340-032-X
1. Brown bear–Juvenile literature. [1. Brown bear. 2. Bears.]
I. Title. II. Series: Northern Trek (Mankato, Minn.)

QL737.C27 G57   2000
599.784–dc21   99-048770
First Edition

2   4   6   8   9   7   5   3   1

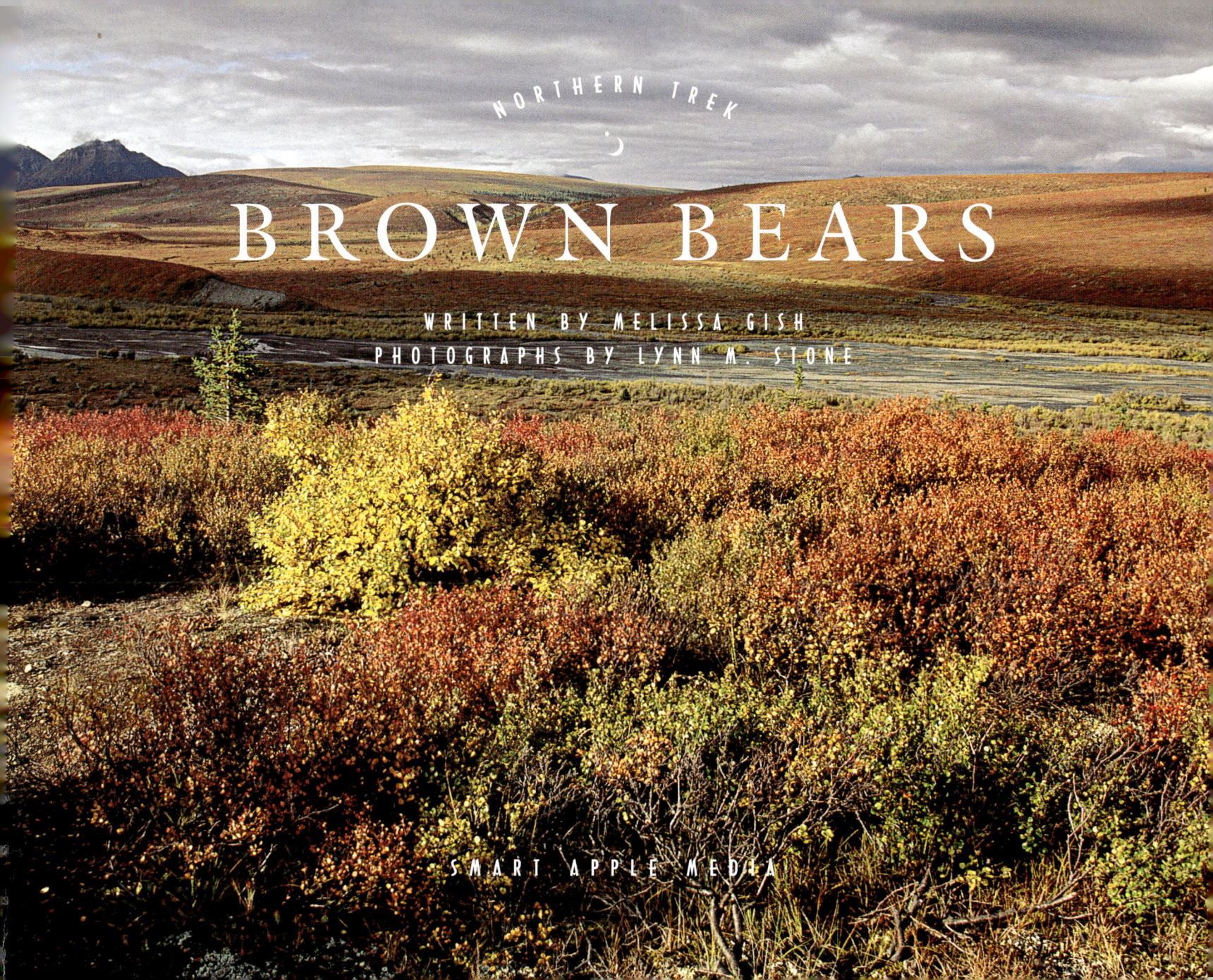

NORTHERN TREK

BROWN BEARS

WRITTEN BY MELISSA GISH
PHOTOGRAPHS BY LYNN M. STONE

SMART APPLE MEDIA

Late September clouds roll across a darkening sky. Cold air

blows down from Denali, the mountain also named McKinley.

A hulking 1,500-pound (680 kg) animal begins its climb up the

mountain: an Alaskan brown bear, the largest land carnivore on Earth.

Trekking toward a mountainside den, where it will sleep through the

winter, the bear takes long strides. It steps into the very same tracks it

stepped into the year before, and the year before—tracks made generations

ago by bears now long since gone. No one knows why bears do this.

These ancient trails are one of the mysteries of nature.

BEARS BELONG TO

a family of mammals called *Ursidae*. The brown bear is one of the two true species of North American forest and mountain bears. The other is the black bear. The brown bear is found only on the western edge of Canada and the United States. Brown bears found on Kodiak Island in the Aleutian Islands around Alaska's southern coast are also called Kodiak bears. The brown bear species also includes the subspecies grizzly, a slightly smaller bear that is most often found in dense interior forests.

Male brown bears stand up to 10 feet (3 m) tall on their hind legs; the largest one ever recorded weighed 1,650 pounds (749 kg). Females are about half this size. Though their name suggests they are all brown, these bears can actually be cream colored, gray, red, dark brown, or black. Older bears tend to be darker than younger bears.

Bears are intelligent, curious creatures that have adapted well to life in the cold northland.

At certain times of the year, bears may be playful and social, but usually they are independent and short-tempered.

6

Because they walk flat on their feet (like humans), they are agile and can run quite fast—up to 35 miles (56 km) per hour—to hunt down swift animals such as elk and deer. Their powerful paws, each with five **retractable** claws, can kill an adult deer with one swipe to the neck.

The bear's sense of smell is keener than that of humans. Bears use scents to find food, select mates, and locate enemies. Its eyesight is similar to that of humans. Also, like humans, bears are **omnivores**. They eat both plants and animals, so they have two kinds of teeth. Molars, the flat teeth at the back of the jaw, grind up plants, fruits, and nuts. The sharp, pointed teeth at the front of the jaw are

---

Bears are not ferocious hunters; in fact, they seldom attack large prey. Rather, they prefer to catch fish, graze, and dig roots and tubers out of the earth.

Bears are related to dogs, so their footprints resemble those of dogs. The bones, however, are much thicker to support the bear's weight, and the claws can grow up to five inches (13 cm) in length.

called canines. These are used to tear up meat.

Bears have such powerful jaws that they can crush the spine of a prey animal with one bite. They hunt both small animals (such as marmots and ground squirrels) and large ones (including caribou and moose). Also, because bears are naturally scavengers, they will eat **carrion** whenever available.

Brown bears don't always hunt for their food, though. Plants and berries make up over half of the animal's diet. A bear may **graze** for hours on the

While brown bears prefer living the solitary life, they will often tolerate each other at riverbanks during periods of heavy fishing.

A brown bear's territory, called its "home range," is where the bear lives and hunts. Home ranges average 30 to 50 square miles (78–130 km²), though studies have reported a few up to 270 square miles (700 km²).

leaves, stems, and roots of wildflowers and grasses found in its territory.

Territories of males and females may overlap, as males will **breed** with more than one female. Two males, however, will not share a territory. They will fight until one of them leaves the area.

Males also fight with each other over the right to mate with the females. After mating takes place, in May and June, male and female partners go off to live by themselves the rest of the year. In fact, bears tend to avoid contact with each other altogether as they spend the warmer months hunting and eating.

In preparation for the coming winter, when they will need to sleep, bears do almost nothing all summer but eat. Because bears are good swimmers

with thick protective fur, they can stand in icy rivers and catch fish all day. This becomes a major part of the brown bear's feeding ritual each year, as millions of Pacific salmon travel from the oceans back to the northern streams where they were hatched. There they lay eggs of their own, a behavior called spawning.

For the brown bear, salmon spawning is an important event, a time when they will eat almost constantly. As many as 40 bears may inhabit a one-mile (1.6 km) stretch of shoreline to catch the salmon as they journey upstream. Each day, a bear will catch and eat about 30 salmon, gaining two to

Whenever brown bears fight over territory, it's rarely deadly. One of them will almost always give up and move on before causing–or receiving–serious damage.

15

three pounds (.9–1.3 kg) of body fat. In one season, a bear may gain as much as 200 pounds (91 kg), adding a six-inch (15 cm) layer of fat to its body.

After the summer feeding is over, all bears travel to dens to sleep. Those females that will give birth awaken in late January or early February.

Normally, one or two cubs are born, though some litters may contain up to four. The female will then continue sleeping while her young **nurse** throughout the winter.

Bear cubs are born blind and toothless; they are about the size of rats, but they grow very quickly their first year.

Brown bears experience long periods of heavy sleep, but they do not truly hibernate. Though they do not eat, drink, or eliminate body waste for about six months (from late fall until spring) they can wake up whenever they want.

A bear cub weighs about one pound (.45 kg) and is born with its eyes closed. Its fine fur makes the cub look hairless. At three weeks old, the cub's eyes open and it may begin to move around the den.

In late March, brown bears start to wake up and leave their dens. The males are the first to leave. For safety, females with new cubs stick close to their dens for several weeks. It's common for adult males to kill and eat small cubs, so the females must keep their offspring close by. A female will fight to the death in an attempt to save her cubs.

At two to three years of age, cubs are chased away by their mother, as she will be getting ready to mate again and have a new litter of cubs. This is another dangerous time in a young bear's life.

Often, young bears—those less than five years old— fall prey to large adults who chase them away from hunting and fishing grounds or kill them.

As adults, however, the brown bear's only enemies are humans. But less than half of all brown bear cubs born survive to adulthood. This is why conservation is so vital to the survival of the Alaskan brown bear.

In 1941 President Franklin Roosevelt estab-

lished the Kodiak Island National Wildlife Refuge, setting aside two million acres (809,400 hectares) of **habitat** for brown bears and other Alaskan wildlife. But today, as neighboring forests are destroyed by industry, and as more people move into bear habitats, the future of the brown bear is uncertain. Only an understanding and appreciation of this magnificent animal can possibly protect it for future generations.

Alaskan and Canadian tribes of Native Americans respected the powerful brown bear and lived in peace with this animal for centuries. When protected from human interference, brown bears may live to be 30 years old.

VIEWING AREAS

MANY PLACES PROVIDE fairly close views of brown bears in the wild, but they are highly popular and visitors often must reserve campsites or sign up for tours well in advance. Guided hikes or boating tours are the best way to enter bear habitats safely, and often young children are discouraged from embarking on the hikes, as many are over dangerously rugged terrain. Listed here are some brown bear habitats with public access. As with any trek into nature, it is important to remember that wild animals are unpredictable and can be dangerous if approached. The best way to view wildlife is from a respectful—and safe—distance.

DENALI NATIONAL PARK IN ALASKA *Here is where the highest mountain in North America, Mt. McKinley, is found, and where many brown bears sleep through the winter months. In the spring and fall, brown bears can be seen trekking up and down the mountainside. Knowledgeable rangers lead Discovery Hikes and give sled dog demonstrations.*

KATMAI NATIONAL PARK IN ALASKA *Visitors to the Brooks River can view brown bears during their annual feeding on sockeye salmon. Two bear-viewing platforms are located along the river, and interpretive rangers lead hikes to an archaeological site and give slide presentations on the area's fascinating background.*

KODIAK ISLAND NATIONAL PARK *Only boats and planes can carry visitors to this island, which is home to about 3,000 brown bears. This is the perfect environment for watching bears in their natural habitat. Though there has never been a human killed by a brown bear on Kodiak Island, campers and fishermen must always be on the lookout for bears seeking food from campsites.*

GLOSSARY & INDEX

breed: *when a male and female animal mate to produce offspring*

carrion: *the body of a dead animal*

graze: *to roam and randomly eat grasses, flowers, or other simple foods*

habitat: *a place where a plant or animal normally lives*

nurse: *to drink milk from a mother animal*

omnivores: *those animals that eat both plants and animal meat*

retractable: *able to be pulled inside something, such as a paw*

black bear, 6
breeding, 15
coats, 6
conservation, 19
cubs, 17–19
eyesight, 9
feeding, 9, 10, 15, 17
grizzly, 6
habitat, 6, 19
hunting, 10, 15

Native Americans, 21
paws, 9
population, 18–19
running, 9
sense of smell, 9
size, 4, 6, 17, 18
sleep, 15, 17, 18
teeth, 9–10
territory, 13, 15
viewing areas, 23